Practice

Eureka Math®
Grade 3 Fluency
Modules 3 & 4

TEKS EDITION

Great Minds® is the creator of *Eureka Math*®, *Wit & Wisdom*®, *Alexandria Plan*™, and *PhD Science*®.

Published by Great Minds PBC
greatminds.org

Printed in the USA

1 2 3 4 5 6 7 8 9 10 CCR 25 24 23 22

ISBN 978-1-63642-856-7

Learn • Practice • Succeed

Eureka Math® student materials for *A Story of Units®* (K–5) are available in the *Learn, Practice, Succeed* trio. This series supports differentiation and remediation while keeping student materials organized and accessible. Educators will find that the *Learn, Practice,* and *Succeed* series also offers coherent—and therefore, more effective—resources for Response to Intervention (RTI), extra practice, and summer learning.

Learn

Eureka Math Learn serves as a student's in-class companion where they show their thinking, share what they know, and watch their knowledge build every day. *Learn* assembles the daily classwork—Application Problems, Exit Tickets, Problem Sets, templates—in an easily stored and navigated volume.

Practice

Each *Eureka Math* lesson begins with a series of energetic, joyous fluency activities, including those found in *Eureka Math Practice*. Students who are fluent in their math facts can master more material more deeply. With *Practice,* students build competence in newly acquired skills and reinforce previous learning in preparation for the next lesson.

Together, *Learn* and *Practice* provide all the print materials students will use for their core math instruction.

Succeed

Eureka Math Succeed enables students to work individually toward mastery. These additional problem sets align lesson by lesson with classroom instruction, making them ideal for use as homework or extra practice. Each problem set is accompanied by a Homework Helper, a set of worked examples that illustrate how to solve similar problems.

Teachers and tutors can use *Succeed* books from prior grade levels as curriculum-consistent tools for filling gaps in foundational knowledge. Students will thrive and progress more quickly as familiar models facilitate connections to their current grade-level content.

Students, families, and educators:

Thank you for being part of the *Eureka Math®* community, where we celebrate the joy, wonder, and thrill of mathematics. One of the most obvious ways we display our excitement is through the fluency activities provided in *Eureka Math Practice*.

What is fluency in mathematics?

You may think of *fluency* as associated with the language arts, where it refers to speaking and writing with ease. In prekindergarten through grade 5, the *Eureka Math* curriculum contains multiple daily opportunities to build fluency *in mathematics*. Each is designed with the same notion—growing every student's ability to use mathematics *with ease*. Fluency experiences are generally fast-paced and energetic, celebrating improvement and focusing on recognizing patterns and connections within the material. They are not intended to be graded.

Eureka Math fluency activities provide differentiated practice through a variety of formats—some are conducted orally, some use manipulatives, others use a personal whiteboard, and still others use a handout and paper-and-pencil format. *Eureka Math Practice* provides each student with the printed fluency exercises for his or her grade level.

What is a Sprint?

Many printed fluency activities utilize the format we call a Sprint. These exercises build speed and accuracy with already acquired skills. Used when students are nearing optimum proficiency, Sprints leverage tempo to build a low-stakes adrenaline boost that increases memory and recall. Their intentional design makes Sprints inherently differentiated; the problems build from simple to complex, with the first quadrant of problems being the simplest and each subsequent quadrant adding complexity. Further, intentional patterns within the sequence of problems engage students' higher order thinking skills.

The suggested format for delivering a Sprint calls for students to do two consecutive Sprints (labeled A and B) on the same skill, each timed at one minute. Students pause between Sprints to articulate the patterns they noticed as they worked the first Sprint. Noticing the patterns often provides a natural boost to their performance on the second Sprint.

Sprints can be conducted with an untimed protocol as well. The untimed protocol is highly recommended when students are still building confidence with the level of complexity of the first quadrant of problems. Once all students are prepared for success on the Sprint, the work of improving speed and accuracy with the energy of a timed protocol is often welcome and invigorating.

Where can I find other fluency activities?

The *Eureka Math Teacher Edition* guides educators in the delivery of all fluency activities for each lesson, including those that do not require print materials. Additionally, the *Eureka Digital Suite* provides access to the fluency activities for all grade levels, searchable by standard or lesson.

Best wishes for a year filled with aha moments!

Jill Diniz

Jill Diniz
Director of Mathematics
Great Minds

Contents

Module 3

Module 4

Grade 3
Module 3

A

Number Correct: _____

Mixed Multiplication

1.	2 × 1 =	
2.	2 × 2 =	
3.	2 × 3 =	
4.	4 × 1 =	
5.	4 × 2 =	
6.	4 × 3 =	
7.	1 × 6 =	
8.	2 × 6 =	
9.	1 × 8 =	
10.	2 × 8 =	
11.	3 × 1 =	
12.	3 × 2 =	
13.	3 × 3 =	
14.	5 × 1 =	
15.	5 × 2 =	
16.	5 × 3 =	
17.	1 × 7 =	
18.	2 × 7 =	
19.	1 × 9 =	
20.	2 × 9 =	
21.	2 × 5 =	
22.	2 × 6 =	

23.	2 × 7 =	
24.	5 × 5 =	
25.	5 × 6 =	
26.	5 × 7 =	
27.	4 × 5 =	
28.	4 × 6 =	
29.	4 × 7 =	
30.	3 × 5 =	
31.	3 × 6 =	
32.	3 × 7 =	
33.	2 × 7 =	
34.	2 × 8 =	
35.	2 × 9 =	
36.	5 × 7 =	
37.	5 × 8 =	
38.	5 × 9 =	
39.	4 × 7 =	
40.	4 × 8 =	
41.	4 × 9 =	
42.	3 × 7 =	
43.	3 × 8 =	
44.	3 × 9 =	

EUREKA MATH®
TEKS EDITION

Lesson 5: Study commutativity to find known facts of 6, 7, 8, and 9.

B

Number Correct: _____

Improvement: _____

Mixed Multiplication

1.	5 × 1 =		23.	5 × 7 =	
2.	5 × 2 =		24.	2 × 5 =	
3.	5 × 3 =		25.	2 × 6 =	
4.	3 × 1 =		26.	2 × 7 =	
5.	3 × 2 =		27.	3 × 5 =	
6.	3 × 3 =		28.	3 × 6 =	
7.	1 × 7 =		29.	3 × 7 =	
8.	2 × 7 =		30.	4 × 5 =	
9.	1 × 9 =		31.	4 × 6 =	
10.	2 × 9 =		32.	4 × 7 =	
11.	2 × 1 =		33.	5 × 7 =	
12.	2 × 2 =		34.	5 × 8 =	
13.	2 × 3 =		35.	5 × 9 =	
14.	4 × 1 =		36.	2 × 7 =	
15.	4 × 2 =		37.	2 × 8 =	
16.	4 × 3 =		38.	2 × 9 =	
17.	1 × 6 =		39.	3 × 7 =	
18.	2 × 6 =		40.	3 × 8 =	
19.	1 × 8 =		41.	3 × 9 =	
20.	2 × 8 =		42.	4 × 7 =	
21.	5 × 5 =		43.	4 × 8 =	
22.	5 × 6 =		44.	4 × 9 =	

Lesson 5:   Study commutativity to find known facts of 6, 7, 8, and 9.

A

Number Correct: _____

Use the Commutative Property to Multiply

1.	2 × 2 =	
2.	2 × 3 =	
3.	3 × 2 =	
4.	2 × 4 =	
5.	4 × 2 =	
6.	2 × 5 =	
7.	5 × 2 =	
8.	2 × 6 =	
9.	6 × 2 =	
10.	2 × 7 =	
11.	7 × 2 =	
12.	2 × 8 =	
13.	8 × 2 =	
14.	2 × 9 =	
15.	9 × 2 =	
16.	2 × 10 =	
17.	10 × 2 =	
18.	5 × 3 =	
19.	3 × 5 =	
20.	5 × 4 =	
21.	4 × 5 =	
22.	5 × 5 =	

23.	5 × 6 =	
24.	6 × 5 =	
25.	5 × 7 =	
26.	7 × 5 =	
27.	5 × 8 =	
28.	8 × 5 =	
29.	5 × 9 =	
30.	9 × 5 =	
31.	5 × 10 =	
32.	10 × 5 =	
33.	3 × 3 =	
34.	3 × 4 =	
35.	4 × 3 =	
36.	3 × 6 =	
37.	6 × 3 =	
38.	3 × 7 =	
39.	7 × 3 =	
40.	3 × 8 =	
41.	8 × 3 =	
42.	3 × 9 =	
43.	9 × 3 =	
44.	4 × 4 =	

Lesson 6: Apply the distributive and commutative properties to relate multiplication facts $5 × n + n$ to $6 × n$ and $n × 6$ where n is the size of the unit.

© Great Minds PBC TEKS Edition | greatminds.org/Texas

7

B

Number Correct: _____

Use the Commutative Property to Multiply

Improvement: _____

1.	$5 \times 2 =$	
2.	$2 \times 5 =$	
3.	$5 \times 3 =$	
4.	$3 \times 5 =$	
5.	$5 \times 4 =$	
6.	$4 \times 5 =$	
7.	$5 \times 5 =$	
8.	$5 \times 6 =$	
9.	$6 \times 5 =$	
10.	$5 \times 7 =$	
11.	$7 \times 5 =$	
12.	$5 \times 8 =$	
13.	$8 \times 5 =$	
14.	$5 \times 9 =$	
15.	$9 \times 5 =$	
16.	$5 \times 10 =$	
17.	$10 \times 5 =$	
18.	$2 \times 2 =$	
19.	$2 \times 3 =$	
20.	$3 \times 2 =$	
21.	$2 \times 4 =$	
22.	$4 \times 2 =$	

23.	$6 \times 2 =$	
24.	$2 \times 6 =$	
25.	$2 \times 7 =$	
26.	$7 \times 2 =$	
27.	$2 \times 8 =$	
28.	$8 \times 2 =$	
29.	$2 \times 9 =$	
30.	$9 \times 2 =$	
31.	$2 \times 10 =$	
32.	$10 \times 2 =$	
33.	$3 \times 3 =$	
34.	$3 \times 4 =$	
35.	$4 \times 3 =$	
36.	$3 \times 6 =$	
37.	$6 \times 3 =$	
38.	$3 \times 7 =$	
39.	$7 \times 3 =$	
40.	$3 \times 8 =$	
41.	$8 \times 3 =$	
42.	$3 \times 9 =$	
43.	$9 \times 3 =$	
44.	$4 \times 4 =$	

Lesson 6:   Apply the distributive and commutative properties to relate multiplication facts $5 \times n + n$ to $6 \times n$ and $n \times 6$ where n is the size of the unit.

© Great Minds PBC TEKS Edition |
greatminds.org/Texas

9

Multiply.

5 × 1 = _____	5 × 2 = _____	5 × 3 = _____	5 × 4 = _____
5 × 5 = _____	5 × 6 = _____	5 × 7 = _____	5 × 8 = _____
5 × 9 = _____	5 × 10 = _____	5 × 5 = _____	5 × 6 = _____
5 × 5 = _____	5 × 7 = _____	5 × 5 = _____	5 × 8 = _____
5 × 5 = _____	5 × 9 = _____	5 × 5 = _____	5 × 10 = _____
5 × 6 = _____	5 × 5 = _____	5 × 6 = _____	5 × 7 = _____
5 × 6 = _____	5 × 8 = _____	5 × 6 = _____	5 × 9 = _____
5 × 6 = _____	5 × 7 = _____	5 × 6 = _____	5 × 7 = _____
5 × 8 = _____	5 × 7 = _____	5 × 9 = _____	5 × 7 = _____
5 × 8 = _____	5 × 6 = _____	5 × 8 = _____	5 × 7 = _____
5 × 8 = _____	5 × 9 = _____	5 × 9 = _____	5 × 6 = _____
5 × 9 = _____	5 × 7 = _____	5 × 9 = _____	5 × 8 = _____
5 × 9 = _____	5 × 8 = _____	5 × 6 = _____	5 × 9 = _____
5 × 7 = _____	5 × 9 = _____	5 × 6 = _____	5 × 8 = _____
5 × 9 = _____	5 × 7 = _____	5 × 6 = _____	5 × 8 = _____

multiply by 5 (6–10)

Lesson 7: Multiply and divide with familiar facts using a box to represent the unknown.

© Great Minds PBC TEKS Edition |
greatminds.org/Texas

Multiply.

6 × 1 = _____	6 × 2 = _____	6 × 3 = _____	6 × 4 = _____
6 × 5 = _____	6 × 1 = _____	6 × 2 = _____	6 × 1 = _____
6 × 3 = _____	6 × 1 = _____	6 × 4 = _____	6 × 1 = _____
6 × 5 = _____	6 × 1 = _____	6 × 2 = _____	6 × 3 = _____
6 × 2 = _____	6 × 4 = _____	6 × 2 = _____	6 × 5 = _____
6 × 2 = _____	6 × 1 = _____	6 × 2 = _____	6 × 3 = _____
6 × 1 = _____	6 × 3 = _____	6 × 2 = _____	6 × 3 = _____
6 × 4 = _____	6 × 3 = _____	6 × 5 = _____	6 × 3 = _____
6 × 4 = _____	6 × 1 = _____	6 × 4 = _____	6 × 2 = _____
6 × 4 = _____	6 × 3 = _____	6 × 4 = _____	6 × 5 = _____
6 × 4 = _____	6 × 5 = _____	6 × 1 = _____	6 × 5 = _____
6 × 2 = _____	6 × 5 = _____	6 × 3 = _____	6 × 5 = _____
6 × 4 = _____	6 × 2 = _____	6 × 4 = _____	6 × 3 = _____
6 × 5 = _____	6 × 3 = _____	6 × 2 = _____	6 × 4 = _____
6 × 3 = _____	6 × 5 = _____	6 × 2 = _____	6 × 4 = _____

multiply by 6 (1–5)

Lesson 9:   Count by units of 7 to multiply and divide using number bonds to decompose.

© Great Minds PBC TEKS Edition | greatminds.org/Texas

Multiply.

6 × 1 = _____   6 × 2 = _____   6 × 3 = _____   6 × 4 = _____

6 × 5 = _____   6 × 6 = _____   6 × 7 = _____   6 × 8 = _____

6 × 9 = _____   6 × 10 = _____   6 × 5 = _____   6 × 6 = _____

6 × 5 = _____   6 × 7 = _____   6 × 5 = _____   6 × 8 = _____

6 × 5 = _____   6 × 9 = _____   6 × 5 = _____   6 × 10 = _____

6 × 6 = _____   6 × 5 = _____   6 × 6 = _____   6 × 7 = _____

6 × 6 = _____   6 × 8 = _____   6 × 6 = _____   6 × 9 = _____

6 × 6 = _____   6 × 7 = _____   6 × 6 = _____   6 × 7 = _____

6 × 8 = _____   6 × 7 = _____   6 × 9 = _____   6 × 7 = _____

6 × 8 = _____   6 × 6 = _____   6 × 8 = _____   6 × 7 = _____

6 × 8 = _____   6 × 9 = _____   6 × 9 = _____   6 × 6 = _____

6 × 9 = _____   6 × 7 = _____   6 × 9 = _____   6 × 8 = _____

6 × 9 = _____   6 × 8 = _____   6 × 6 = _____   6 × 9 = _____

6 × 7 = _____   6 × 9 = _____   6 × 6 = _____   6 × 8 = _____

6 × 9 = _____   6 × 7 = _____   6 × 6 = _____   6 × 8 = _____

multiply by 6 (6–10)

EUREKA MATH TEKS EDITION

Lesson 10: Interpret the unknown in multiplication and division to model and solve problems using units of 6 and 7.

© Great Minds PBC TEKS Edition | greatminds.org/Texas

15

Multiply.

7 × 1 = _____	7 × 2 = _____	7 × 3 = _____	7 × 4 = _____
7 × 5 = _____	7 × 1 = _____	7 × 2 = _____	7 × 1 = _____
7 × 3 = _____	7 × 1 = _____	7 × 4 = _____	7 × 1 = _____
7 × 5 = _____	7 × 1 = _____	7 × 2 = _____	7 × 3 = _____
7 × 2 = _____	7 × 4 = _____	7 × 2 = _____	7 × 5 = _____
7 × 2 = _____	7 × 1 = _____	7 × 2 = _____	7 × 3 = _____
7 × 1 = _____	7 × 3 = _____	7 × 2 = _____	7 × 3 = _____
7 × 4 = _____	7 × 3 = _____	7 × 5 = _____	7 × 3 = _____
7 × 4 = _____	7 × 1 = _____	7 × 4 = _____	7 × 2 = _____
7 × 4 = _____	7 × 3 = _____	7 × 4 = _____	7 × 5 = _____
7 × 4 = _____	7 × 5 = _____	7 × 1 = _____	7 × 5 = _____
7 × 2 = _____	7 × 5 = _____	7 × 3 = _____	7 × 5 = _____
7 × 4 = _____	7 × 2 = _____	7 × 4 = _____	7 × 3 = _____
7 × 5 = _____	7 × 3 = _____	7 × 2 = _____	7 × 4 = _____
7 × 3 = _____	7 × 5 = _____	7 × 2 = _____	7 × 4 = _____

_____

multiply by 7 (1–5)

Lesson 11:   Understand the function of parentheses and apply to solving problems.

17

Multiply.

7 × 1 = _____	7 × 2 = _____	7 × 3 = _____	7 × 4 = _____
7 × 5 = _____	7 × 6 = _____	7 × 7 = _____	7 × 8 = _____
7 × 9 = _____	7 × 10 = _____	7 × 5 = _____	7 × 6 = _____
7 × 5 = _____	7 × 7 = _____	7 × 5 = _____	7 × 8 = _____
7 × 5 = _____	7 × 9 = _____	7 × 5 = _____	7 × 10 = _____
7 × 6 = _____	7 × 5 = _____	7 × 6 = _____	7 × 7 = _____
7 × 6 = _____	7 × 8 = _____	7 × 6 = _____	7 × 9 = _____
7 × 6 = _____	7 × 7 = _____	7 × 6 = _____	7 × 7 = _____
7 × 8 = _____	7 × 7 = _____	7 × 9 = _____	7 × 7 = _____
7 × 8 = _____	7 × 6 = _____	7 × 8 = _____	7 × 7 = _____
7 × 8 = _____	7 × 9 = _____	7 × 9 = _____	7 × 6 = _____
7 × 9 = _____	7 × 7 = _____	7 × 9 = _____	7 × 8 = _____
7 × 9 = _____	7 × 8 = _____	7 × 6 = _____	7 × 9 = _____
7 × 7 = _____	7 × 9 = _____	7 × 6 = _____	7 × 8 = _____
7 × 9 = _____	7 × 7 = _____	7 × 6 = _____	7 × 8 = _____

multiply by 7 (6–10)

Lesson 12:   Model the associative property as a strategy to multiply.

19

Multiply.

8 × 1 = _____   8 × 2 = _____   8 × 3 = _____   8 × 4 = _____

8 × 5 = _____   8 × 1 = _____   8 × 2 = _____   8 × 1 = _____

8 × 3 = _____   8 × 1 = _____   8 × 4 = _____   8 × 1 = _____

8 × 5 = _____   8 × 1 = _____   8 × 2 = _____   8 × 3 = _____

8 × 2 = _____   8 × 4 = _____   8 × 2 = _____   8 × 5 = _____

8 × 2 = _____   8 × 1 = _____   8 × 2 = _____   8 × 3 = _____

8 × 1 = _____   8 × 3 = _____   8 × 2 = _____   8 × 3 = _____

8 × 4 = _____   8 × 3 = _____   8 × 5 = _____   8 × 3 = _____

8 × 4 = _____   8 × 1 = _____   8 × 4 = _____   8 × 2 = _____

8 × 4 = _____   8 × 3 = _____   8 × 4 = _____   8 × 5 = _____

8 × 4 = _____   8 × 5 = _____   8 × 1 = _____   8 × 5 = _____

8 × 2 = _____   8 × 5 = _____   8 × 3 = _____   8 × 5 = _____

8 × 4 = _____   8 × 2 = _____   8 × 4 = _____   8 × 3 = _____

8 × 5 = _____   8 × 3 = _____   8 × 2 = _____   8 × 4 = _____

8 × 3 = _____   8 × 5 = _____   8 × 2 = _____   8 × 4 = _____

multiply by 8 (1–5)

Lesson 13:   Apply the distributive property and the fact 9 = 10 − 1 as a strategy
to multiply.

Multiply.

8 × 1 = _____   8 × 2 = _____   8 × 3 = _____   8 × 4 = _____

8 × 5 = _____   8 × 6 = _____   8 × 7 = _____   8 × 8 = _____

8 × 9 = _____   8 × 10 = _____   8 × 5 = _____   8 × 6 = _____

8 × 5 = _____   8 × 7 = _____   8 × 5 = _____   8 × 8 = _____

8 × 5 = _____   8 × 9 = _____   8 × 5 = _____   8 × 10 = _____

8 × 6 = _____   8 × 5 = _____   8 × 6 = _____   8 × 7 = _____

8 × 6 = _____   8 × 8 = _____   8 × 6 = _____   8 × 9 = _____

8 × 6 = _____   8 × 7 = _____   8 × 6 = _____   8 × 7 = _____

8 × 8 = _____   8 × 7 = _____   8 × 9 = _____   8 × 7 = _____

8 × 8 = _____   8 × 6 = _____   8 × 8 = _____   8 × 7 = _____

8 × 8 = _____   8 × 9 = _____   8 × 9 = _____   8 × 6 = _____

8 × 9 = _____   8 × 7 = _____   8 × 9 = _____   8 × 8 = _____

8 × 9 = _____   8 × 8 = _____   8 × 6 = _____   8 × 9 = _____

8 × 7 = _____   8 × 9 = _____   8 × 6 = _____   8 × 8 = _____

8 × 9 = _____   8 × 7 = _____   8 × 6 = _____   8 × 8 = _____

multiply by 8 (6–10)

Lesson 14:   Interpret the unknown in multiplication and division to model and solve problems.

© Great Minds PBC TEKS Edition |
greatminds.org/Texas

A

Number Correct: _____

Multiply or divide by 8

1.	$2 \times 8 =$	
2.	$3 \times 8 =$	
3.	$4 \times 8 =$	
4.	$5 \times 8 =$	
5.	$1 \times 8 =$	
6.	$16 \div 8 =$	
7.	$24 \div 8 =$	
8.	$40 \div 8 =$	
9.	$8 \div 1 =$	
10.	$32 \div 8 =$	
11.	$6 \times 8 =$	
12.	$7 \times 8 =$	
13.	$8 \times 8 =$	
14.	$9 \times 8 =$	
15.	$10 \times 8 =$	
16.	$64 \div 8 =$	
17.	$56 \div 8 =$	
18.	$72 \div 8 =$	
19.	$48 \div 8 =$	
20.	$80 \div 8 =$	
21.	_____ $\times 8 = 40$	
22.	_____ $\times 8 = 16$	

23.	_____ $\times 8 = 80$	
24.	_____ $\times 8 = 32$	
25.	_____ $\times 8 = 24$	
26.	$80 \div 8 =$	
27.	$40 \div 8 =$	
28.	$8 \div 1 =$	
29.	$16 \div 8 =$	
30.	$24 \div 8 =$	
31.	_____ $\times 8 = 48$	
32.	_____ $\times 8 = 56$	
33.	_____ $\times 8 = 72$	
34.	_____ $\times 8 = 64$	
35.	$56 \div 8 =$	
36.	$72 \div 8 =$	
37.	$48 \div 8 =$	
38.	$64 \div 8 =$	
39.	$11 \times 8 =$	
40.	$88 \div 8 =$	
41.	$12 \times 8 =$	
42.	$96 \div 8 =$	
43.	$14 \times 8 =$	
44.	$112 \div 8 =$	

Lesson 15: Reason about and explain arithmetic patterns using units
of 0 and 1 as they relate to multiplication and division.

© Great Minds PBC TEKS Edition |
greatminds.org/Texas

B

Number Correct: _____

Improvement: _____

Multiply or divide by 8

1.	1 × 8 =	
2.	2 × 8 =	
3.	3 × 8 =	
4.	4 × 8 =	
5.	5 × 8 =	
6.	24 ÷ 8 =	
7.	16 ÷ 8 =	
8.	32 ÷ 8 =	
9.	8 ÷ 1 =	
10.	40 ÷ 8 =	
11.	10 × 8 =	
12.	6 × 8 =	
13.	7 × 8 =	
14.	8 × 8 =	
15.	9 × 8 =	
16.	56 ÷ 8 =	
17.	48 ÷ 8 =	
18.	64 ÷ 8 =	
19.	80 ÷ 8 =	
20.	72 ÷ 8 =	
21.	_____ × 8 = 16	
22.	_____ × 8 = 40	

23.	_____ × 8 = 48	
24.	_____ × 8 = 80	
25.	_____ × 8 = 24	
26.	16 ÷ 8 =	
27.	8 ÷ 1 =	
28.	80 ÷ 8 =	
29.	40 ÷ 8 =	
30.	24 ÷ 8 =	
31.	_____ × 8 = 64	
32.	_____ × 8 = 32	
33.	_____ × 8 = 72	
34.	_____ × 8 = 56	
35.	64 ÷ 8 =	
36.	72 ÷ 8 =	
37.	48 ÷ 8 =	
38.	56 ÷ 8 =	
39.	11 × 8 =	
40.	88 ÷ 8 =	
41.	12 × 8 =	
42.	96 ÷ 8 =	
43.	13 × 8 =	
44.	104 ÷ 8 =	

EUREKA MATH
TEKS EDITION

Lesson 15: Reason about and explain arithmetic patterns using units of 0 and 1 as they relate to multiplication and division.

A

Number Correct: _____

Multiply and Divide with 1 and 0

1.	_____ × 1 = 2	
2.	_____ × 1 = 3	
3.	_____ × 1 = 4	
4.	_____ × 1 = 9	
5.	8 × _____ = 0	
6.	9 × _____ = 0	
7.	4 × _____ = 0	
8.	5 × _____ = 5	
9.	6 × _____ = 6	
10.	7 × _____ = 7	
11.	3 × _____ = 3	
12.	0 ÷ 1 = _____	
13.	0 ÷ 2 = _____	
14.	0 ÷ 3 = _____	
15.	0 ÷ 6 = _____	
16.	1 × _____ = 1	
17.	4 ÷ _____ = 4	
18.	5 ÷ _____ = 5	
19.	6 ÷ _____ = 6	
20.	8 ÷ _____ = 8	
21.	_____ × 1 = 5	
22.	3 × _____ = 0	

23.	9 ÷ _____ = 9	
24.	8 × _____ = 8	
25.	_____ × 1 = 1	
26.	0 ÷ 3 = _____	
27.	_____ × 1 = 7	
28.	6 × _____ = 0	
29.	4 × _____ = 4	
30.	0 ÷ 8 = _____	
31.	0 × _____ = 0	
32.	1 ÷ 1 = _____	
33.	_____ × 1 = 24	
34.	17 × _____ = 0	
35.	32 × _____ = 32	
36.	0 ÷ 19 = _____	
37.	46 × _____ = 0	
38.	0 ÷ 51 = _____	
39.	64 × _____ = 64	
40.	_____ × 1 = 79	
41.	0 ÷ 82 = _____	
42.	_____ × 1 = 96	
43.	27 × _____ = 27	
44.	43 × _____ = 0	

Lesson 17: Solve two-step word problems involving all four operations and assess the reasonableness of solutions.

B

Number Correct: _____

Multiply and Divide with 1 and 0

Improvement: _____

1.	_____ × 1 = 3	
2.	_____ × 1 = 4	
3.	_____ × 1 = 5	
4.	_____ × 1 = 8	
5.	7 × _____ = 0	
6.	8 × _____ = 0	
7.	3 × _____ = 0	
8.	4 × _____ = 4	
9.	5 × _____ = 5	
10.	6 × _____ = 6	
11.	2 × _____ = 2	
12.	0 ÷ 2 = _____	
13.	0 ÷ 3 = _____	
14.	0 ÷ 4 = _____	
15.	0 ÷ 7 = _____	
16.	1 × _____ = 1	
17.	3 ÷ _____ = 3	
18.	4 ÷ _____ = 4	
19.	5 ÷ _____ = 5	
20.	7 ÷ _____ = 7	
21.	_____ × 1 = 6	
22.	4 × _____ = 0	

23.	8 ÷ _____ = 8	
24.	7 × _____ = 7	
25.	_____ × 1 = 1	
26.	0 ÷ 5 = _____	
27.	_____ × 1 = 9	
28.	5 × _____ = 0	
29.	9 × _____ = 9	
30.	0 ÷ 6 = _____	
31.	1 ÷ 1 = _____	
32.	0 × _____ = 0	
33.	_____ × 1 = 34	
34.	16 × _____ = 0	
35.	31 × _____ = 31	
36.	0 ÷ 18 = _____	
37.	45 × _____ = 0	
38.	0 ÷ 52 = _____	
39.	63 × _____ = 63	
40.	_____ × 1 = 78	
41.	0 ÷ 81 = _____	
42.	_____ × 1 = 97	
43.	26 × _____ = 26	
44.	42 × _____ = 0	

EUREKA MATH
TEKS EDITION

Lesson 17: Solve two-step word problems involving all four operations and assess the reasonableness of solutions.

© Great Minds PBC TEKS Edition |
greatminds.org/Texas

Multiply.

9 × 1 = _____   9 × 2 = _____   9 × 3 = _____   9 × 4 = _____

9 × 5 = _____   9 × 1 = _____   9 × 2 = _____   9 × 1 = _____

9 × 3 = _____   9 × 1 = _____   9 × 4 = _____   9 × 1 = _____

9 × 5 = _____   9 × 1 = _____   9 × 2 = _____   9 × 3 = _____

9 × 2 = _____   9 × 4 = _____   9 × 2 = _____   9 × 5 = _____

9 × 2 = _____   9 × 1 = _____   9 × 2 = _____   9 × 3 = _____

9 × 1 = _____   9 × 3 = _____   9 × 2 = _____   9 × 3 = _____

9 × 4 = _____   9 × 3 = _____   9 × 5 = _____   9 × 3 = _____

9 × 4 = _____   9 × 1 = _____   9 × 4 = _____   9 × 2 = _____

9 × 4 = _____   9 × 3 = _____   9 × 4 = _____   9 × 5 = _____

9 × 4 = _____   9 × 5 = _____   9 × 1 = _____   9 × 5 = _____

9 × 2 = _____   9 × 5 = _____   9 × 3 = _____   9 × 5 = _____

9 × 4 = _____   9 × 2 = _____   9 × 4 = _____   9 × 3 = _____

9 × 5 = _____   9 × 3 = _____   9 × 2 = _____   9 × 4 = _____

9 × 3 = _____   9 × 5 = _____   9 × 2 = _____   9 × 4 = _____

multiply by 9 (1–5)

Lesson 19:   Use place value strategies and the associative property
$n \times (m \times 10) = (n \times m) \times 10$ (where n and m are less than 10) to
multiply by multiples of 10.

A

Number Correct: _____

Multiply by Multiples of 10

1.	$2 \times 3 =$	
2.	$2 \times 30 =$	
3.	$20 \times 3 =$	
4.	$2 \times 2 =$	
5.	$2 \times 20 =$	
6.	$20 \times 2 =$	
7.	$4 \times 2 =$	
8.	$4 \times 20 =$	
9.	$40 \times 2 =$	
10.	$5 \times 3 =$	
11.	$50 \times 3 =$	
12.	$3 \times 50 =$	
13.	$4 \times 4 =$	
14.	$40 \times 4 =$	
15.	$4 \times 40 =$	
16.	$6 \times 3 =$	
17.	$6 \times 30 =$	
18.	$60 \times 3 =$	
19.	$7 \times 5 =$	
20.	$70 \times 5 =$	
21.	$7 \times 50 =$	
22.	$8 \times 4 =$	

23.	$8 \times 40 =$	
24.	$80 \times 4 =$	
25.	$9 \times 6 =$	
26.	$90 \times 6 =$	
27.	$2 \times 5 =$	
28.	$2 \times 50 =$	
29.	$3 \times 90 =$	
30.	$40 \times 7 =$	
31.	$5 \times 40 =$	
32.	$6 \times 60 =$	
33.	$70 \times 6 =$	
34.	$8 \times 70 =$	
35.	$80 \times 6 =$	
36.	$9 \times 70 =$	
37.	$50 \times 6 =$	
38.	$8 \times 80 =$	
39.	$9 \times 80 =$	
40.	$60 \times 8 =$	
41.	$70 \times 7 =$	
42.	$5 \times 80 =$	
43.	$60 \times 9 =$	
44.	$9 \times 90 =$	

Lesson 20:   Use concrete models to represent two-digit by one-digit multiplication.

35

B

Number Correct: _____

Multiply by Multiples of 10

Improvement: _____

1.	4 × 2 =	
2.	4 × 20 =	
3.	40 × 2 =	
4.	3 × 3 =	
5.	3 × 30 =	
6.	30 × 3 =	
7.	3 × 2 =	
8.	3 × 20 =	
9.	30 × 2 =	
10.	5 × 5 =	
11.	50 × 5 =	
12.	5 × 50 =	
13.	4 × 3 =	
14.	40 × 3 =	
15.	4 × 30 =	
16.	7 × 3 =	
17.	7 × 30 =	
18.	70 × 3 =	
19.	6 × 4 =	
20.	60 × 4 =	
21.	6 × 40 =	
22.	9 × 4 =	

23.	9 × 40 =	
24.	90 × 4 =	
25.	8 × 6 =	
26.	80 × 6 =	
27.	5 × 2 =	
28.	5 × 20 =	
29.	3 × 80 =	
30.	40 × 8 =	
31.	4 × 50 =	
32.	8 × 80 =	
33.	90 × 6 =	
34.	6 × 70 =	
35.	60 × 6 =	
36.	7 × 70 =	
37.	60 × 5 =	
38.	6 × 80 =	
39.	7 × 80 =	
40.	80 × 6 =	
41.	90 × 7 =	
42.	8 × 50 =	
43.	80 × 9 =	
44.	7 × 90 =	

Lesson 20: Use concrete models to represent two-digit by one-digit multiplication.

Multiply.

9 × 1 = _____   9 × 2 = _____   9 × 3 = _____   9 × 4 = _____

9 × 5 = _____   9 × 6 = _____   9 × 7 = _____   9 × 8 = _____

9 × 9 = _____   9 × 10 = _____   9 × 5 = _____   9 × 6 = _____

9 × 5 = _____   9 × 7 = _____   9 × 5 = _____   9 × 8 = _____

9 × 5 = _____   9 × 9 = _____   9 × 5 = _____   9 × 10 = _____

9 × 6 = _____   9 × 5 = _____   9 × 6 = _____   9 × 7 = _____

9 × 6 = _____   9 × 8 = _____   9 × 6 = _____   9 × 9 = _____

9 × 6 = _____   9 × 7 = _____   9 × 6 = _____   9 × 7 = _____

9 × 8 = _____   9 × 7 = _____   9 × 9 = _____   9 × 7 = _____

9 × 8 = _____   9 × 6 = _____   9 × 8 = _____   9 × 7 = _____

9 × 8 = _____   9 × 9 = _____   9 × 9 = _____   9 × 6 = _____

9 × 9 = _____   9 × 7 = _____   9 × 9 = _____   9 × 8 = _____

9 × 9 = _____   9 × 8 = _____   9 × 6 = _____   9 × 9 = _____

9 × 7 = _____   9 × 9 = _____   9 × 6 = _____   9 × 8 = _____

9 × 9 = _____   9 × 7 = _____   9 × 6 = _____   9 × 8 = _____

_____

multiply by 9 (6–10)

Lesson 21:   Draw models to represent two-digit by one-digit multiplication.

39

A

Number Correct: _____

Multiply or divide by 9

1.	2 × 9 =	
2.	3 × 9 =	
3.	4 × 9 =	
4.	5 × 9 =	
5.	1 × 9 =	
6.	18 ÷ 9 =	
7.	27 ÷ 9 =	
8.	45 ÷ 9 =	
9.	9 ÷ 9 =	
10.	36 ÷ 9 =	
11.	6 × 9 =	
12.	7 × 9 =	
13.	8 × 9 =	
14.	9 × 9 =	
15.	10 × 9 =	
16.	72 ÷ 9 =	
17.	63 ÷ 9 =	
18.	81 ÷ 9 =	
19.	54 ÷ 9 =	
20.	90 ÷ 9 =	
21.	_____ × 9 = 45	
22.	_____ × 9 = 9	

23.	_____ × 9 = 90	
24.	_____ × 9 = 18	
25.	_____ × 9 = 27	
26.	90 ÷ 9 =	
27.	45 ÷ 9 =	
28.	9 ÷ 9 =	
29.	18 ÷ 9 =	
30.	27 ÷ 9 =	
31.	_____ × 9 = 54	
32.	_____ × 9 = 63	
33.	_____ × 9 = 81	
34.	_____ × 9 = 72	
35.	63 ÷ 9 =	
36.	81 ÷ 9 =	
37.	54 ÷ 9 =	
38.	72 ÷ 9 =	
39.	11 × 9 =	
40.	99 ÷ 9 =	
41.	12 × 9 =	
42.	108 ÷ 9 =	
43.	14 × 9 =	
44.	126 ÷ 9 =	

Lesson 22:   Multiply two-digit numbers by one-digit numbers using the standard algorithm.

B

Number Correct: _____

Improvement: _____

Multiply or divide by 9

1.	$1 \times 9 =$	
2.	$2 \times 9 =$	
3.	$3 \times 9 =$	
4.	$4 \times 9 =$	
5.	$5 \times 9 =$	
6.	$27 \div 9 =$	
7.	$18 \div 9 =$	
8.	$36 \div 9 =$	
9.	$9 \div 9 =$	
10.	$45 \div 9 =$	
11.	$10 \times 9 =$	
12.	$6 \times 9 =$	
13.	$7 \times 9 =$	
14.	$8 \times 9 =$	
15.	$9 \times 9 =$	
16.	$63 \div 9 =$	
17.	$54 \div 9 =$	
18.	$72 \div 9 =$	
19.	$90 \div 9 =$	
20.	$81 \div 9 =$	
21.	_____ $\times 9 = 9$	
22.	_____ $\times 9 = 45$	

23.	_____ $\times 9 = 18$	
24.	_____ $\times 9 = 90$	
25.	_____ $\times 9 = 27$	
26.	$18 \div 9 =$	
27.	$9 \div 9 =$	
28.	$90 \div 9 =$	
29.	$45 \div 9 =$	
30.	$27 \div 9 =$	
31.	_____ $\times 9 = 27$	
32.	_____ $\times 9 = 36$	
33.	_____ $\times 9 = 81$	
34.	_____ $\times 9 = 63$	
35.	$72 \div 9 =$	
36.	$81 \div 9 =$	
37.	$54 \div 9 =$	
38.	$63 \div 9 =$	
39.	$11 \times 9 =$	
40.	$99 \div 9 =$	
41.	$12 \times 9 =$	
42.	$108 \div 9 =$	
43.	$13 \times 9 =$	
44.	$117 \div 9 =$	

Lesson 22: Multiply two-digit numbers by one-digit numbers using the standard algorithm.

43

EUREKA MATH®
TEKS EDITION

© Great Minds PBC TEKS Edition |
greatminds.org/Texas

Grade 3
Module 4

Multiply.

6 × 1 = _____   6 × 2 = _____   6 × 3 = _____   6 × 4 = _____

6 × 5 = _____   6 × 6 = _____   6 × 7 = _____   6 × 8 = _____

6 × 9 = _____   6 × 10 = _____   6 × 5 = _____   6 × 6 = _____

6 × 5 = _____   6 × 7 = _____   6 × 5 = _____   6 × 8 = _____

6 × 5 = _____   6 × 9 = _____   6 × 5 = _____   6 × 10 = _____

6 × 6 = _____   6 × 5 = _____   6 × 6 = _____   6 × 7 = _____

6 × 6 = _____   6 × 8 = _____   6 × 6 = _____   6 × 9 = _____

6 × 6 = _____   6 × 7 = _____   6 × 6 = _____   6 × 7 = _____

6 × 8 = _____   6 × 7 = _____   6 × 9 = _____   6 × 7 = _____

6 × 8 = _____   6 × 6 = _____   6 × 8 = _____   6 × 7 = _____

6 × 8 = _____   6 × 9 = _____   6 × 9 = _____   6 × 6 = _____

6 × 9 = _____   6 × 7 = _____   6 × 9 = _____   6 × 8 = _____

6 × 9 = _____   6 × 8 = _____   6 × 6 = _____   6 × 9 = _____

6 × 7 = _____   6 × 9 = _____   6 × 6 = _____   6 × 8 = _____

6 × 9 = _____   6 × 7 = _____   6 × 6 = _____   6 × 8 = _____

multiply by 6 (6–10)

Lesson 5:   Find the area of a rectangle through multiplication of the side lengths.

EUREKA
MATH
TEKS EDITION

Multiply.

7 × 1 = _____	7 × 2 = _____	7 × 3 = _____	7 × 4 = _____
7 × 5 = _____	7 × 6 = _____	7 × 7 = _____	7 × 8 = _____
7 × 9 = _____	7 × 10 = _____	7 × 5 = _____	7 × 6 = _____
7 × 5 = _____	7 × 7 = _____	7 × 5 = _____	7 × 8 = _____
7 × 5 = _____	7 × 9 = _____	7 × 5 = _____	7 × 10 = _____
7 × 6 = _____	7 × 5 = _____	7 × 6 = _____	7 × 7 = _____
7 × 6 = _____	7 × 8 = _____	7 × 6 = _____	7 × 9 = _____
7 × 6 = _____	7 × 7 = _____	7 × 6 = _____	7 × 7 = _____
7 × 8 = _____	7 × 7 = _____	7 × 9 = _____	7 × 7 = _____
7 × 8 = _____	7 × 6 = _____	7 × 8 = _____	7 × 7 = _____
7 × 8 = _____	7 × 9 = _____	7 × 9 = _____	7 × 6 = _____
7 × 9 = _____	7 × 7 = _____	7 × 9 = _____	7 × 8 = _____
7 × 9 = _____	7 × 8 = _____	7 × 6 = _____	7 × 9 = _____
7 × 7 = _____	7 × 9 = _____	7 × 6 = _____	7 × 8 = _____
7 × 9 = _____	7 × 7 = _____	7 × 6 = _____	7 × 8 = _____

multiply by 7 (6–10)

Lesson 9:   Solve word problems involving area.

49

Multiply.

8 × 1 = _____   8 × 2 = _____   8 × 3 = _____   8 × 4 = _____

8 × 5 = _____   8 × 6 = _____   8 × 7 = _____   8 × 8 = _____

8 × 9 = _____   8 × 10 = _____   8 × 5 = _____   8 × 6 = _____

8 × 5 = _____   8 × 7 = _____   8 × 5 = _____   8 × 8 = _____

8 × 5 = _____   8 × 9 = _____   8 × 5 = _____   8 × 10 = _____

8 × 6 = _____   8 × 5 = _____   8 × 6 = _____   8 × 7 = _____

8 × 6 = _____   8 × 8 = _____   8 × 6 = _____   8 × 9 = _____

8 × 6 = _____   8 × 7 = _____   8 × 6 = _____   8 × 7 = _____

8 × 8 = _____   8 × 7 = _____   8 × 9 = _____   8 × 7 = _____

8 × 8 = _____   8 × 6 = _____   8 × 8 = _____   8 × 7 = _____

8 × 8 = _____   8 × 9 = _____   8 × 9 = _____   8 × 6 = _____

8 × 9 = _____   8 × 7 = _____   8 × 9 = _____   8 × 8 = _____

8 × 9 = _____   8 × 8 = _____   8 × 6 = _____   8 × 9 = _____

8 × 7 = _____   8 × 9 = _____   8 × 6 = _____   8 × 8 = _____

8 × 9 = _____   8 × 7 = _____   8 × 6 = _____   8 × 8 = _____

multiply by 8 (6–10)

Lesson 11:   Find areas by decomposing into rectangles or completing composite figures to form rectangles.

© Great Minds PBC TEKS Edition | greatminds.org/Texas

51

Multiply.

9 × 1 = _____   9 × 2 = _____   9 × 3 = _____   9 × 4 = _____

9 × 5 = _____   9 × 1 = _____   9 × 2 = _____   9 × 1 = _____

9 × 3 = _____   9 × 1 = _____   9 × 4 = _____   9 × 1 = _____

9 × 5 = _____   9 × 1 = _____   9 × 2 = _____   9 × 3 = _____

9 × 2 = _____   9 × 4 = _____   9 × 2 = _____   9 × 5 = _____

9 × 2 = _____   9 × 1 = _____   9 × 2 = _____   9 × 3 = _____

9 × 1 = _____   9 × 3 = _____   9 × 2 = _____   9 × 3 = _____

9 × 4 = _____   9 × 3 = _____   9 × 5 = _____   9 × 3 = _____

9 × 4 = _____   9 × 1 = _____   9 × 4 = _____   9 × 2 = _____

9 × 4 = _____   9 × 3 = _____   9 × 4 = _____   9 × 5 = _____

9 × 4 = _____   9 × 5 = _____   9 × 1 = _____   9 × 5 = _____

9 × 2 = _____   9 × 5 = _____   9 × 3 = _____   9 × 5 = _____

9 × 4 = _____   9 × 2 = _____   9 × 4 = _____   9 × 3 = _____

9 × 5 = _____   9 × 3 = _____   9 × 2 = _____   9 × 4 = _____

9 × 3 = _____   9 × 5 = _____   9 × 2 = _____   9 × 4 = _____

multiply by 9 (1–5)

Lesson 12:   Apply knowledge of area to determine areas of rooms in a given floor plan.

Multiply.

9 × 1 = _____   9 × 2 = _____   9 × 3 = _____   9 × 4 = _____

9 × 5 = _____   9 × 6 = _____   9 × 7 = _____   9 × 8 = _____

9 × 9 = _____   9 × 10 = _____   9 × 5 = _____   9 × 6 = _____

9 × 5 = _____   9 × 7 = _____   9 × 5 = _____   9 × 8 = _____

9 × 5 = _____   9 × 9 = _____   9 × 5 = _____   9 × 10 = _____

9 × 6 = _____   9 × 5 = _____   9 × 6 = _____   9 × 7 = _____

9 × 6 = _____   9 × 8 = _____   9 × 6 = _____   9 × 9 = _____

9 × 6 = _____   9 × 7 = _____   9 × 6 = _____   9 × 7 = _____

9 × 8 = _____   9 × 7 = _____   9 × 9 = _____   9 × 7 = _____

9 × 8 = _____   9 × 6 = _____   9 × 8 = _____   9 × 7 = _____

9 × 8 = _____   9 × 9 = _____   9 × 9 = _____   9 × 6 = _____

9 × 9 = _____   9 × 7 = _____   9 × 9 = _____   9 × 8 = _____

9 × 9 = _____   9 × 8 = _____   9 × 6 = _____   9 × 9 = _____

9 × 7 = _____   9 × 9 = _____   9 × 6 = _____   9 × 8 = _____

9 × 9 = _____   9 × 7 = _____   9 × 6 = _____   9 × 8 = _____

multiply by 9 (6–10)

Lesson 13:   Apply knowledge of area to determine areas of rooms in a given floor plan.

Credits

Great Minds® has made every effort to obtain permission for the reprinting of all copyrighted material. If any owner of copyrighted material is not acknowledged herein, please contact Great Minds for proper acknowledgment in all future editions and reprints of these modules.